Problem Solving and Reasoning Pupil Book 4

Peter Clarke

William Collins' dream of knowledge for all began with the publication of his first book in 1819. A self-educated mill worker, he not only enriched millions of lives, but also founded a flourishing publishing house. Today, staying true to this spirit, Collins books are packed with inspiration, innovation and practical expertise. They place you at the centre of a world of possibility and give you exactly what you need to explore it.

Collins. Freedom to teach.

Published by Collins
An imprint of HarperCollins*Publishers*
The News Building
1 London Bridge Street
London
SE1 9GF

HarperCollins*Publishers* Macken House
39/40 Mayor Street Upper
Dublin1
D01 C9W8
Ireland

Browse the complete Collins catalogue at
www.collins.co.uk

10 9 8 7 6 5

ISBN 978-0-00-826049-1

The author asserts his moral rights to be identified as the author of this work.

The author wishes to thank Brian Molyneaux for his valuable contribution to this publication.

British Library Cataloguing in Publication Data
A Catalogue record for this publication is available from the British Library

Author: Peter Clarke
Publishing manager: Fiona McGlade
Editor: Amy Wright
Copyeditor: Catherine Dakin
Proofreader: Tanya Solomons
Answer checker: Steven Matchett
Cover designer: Amparo Barrera
Internal designer: 2hoots Publishing Services
Typesetter: Ken Vail Graphic Design
Illustrator: Eva Sassin
Production controller: Sarah Burke
Printed and bound by Ashford Colour Press Ltd

Contents

Reasoning mathematically

Contents

5

How to use this book

Aims

This book aims to provide teachers with a resource that enables pupils to:

- develop mathematical problem solving and thinking skills
- reason and communicate mathematically
- use and apply mathematics to solve problems.

The three different types of mathematical problem solving challenge

This book consists of three different types of mathematical problem solving challenge:

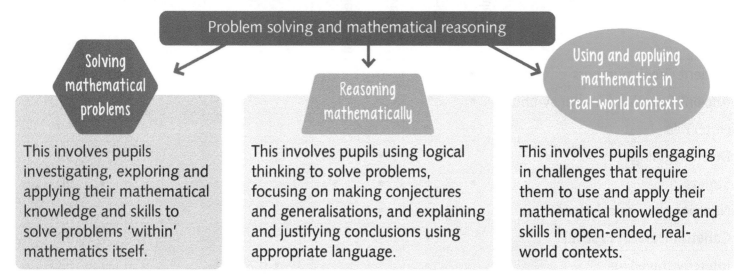

Problem solving and mathematical reasoning

Solving mathematical problems

This involves pupils investigating, exploring and applying their mathematical knowledge and skills to solve problems 'within' mathematics itself.

Reasoning mathematically

This involves pupils using logical thinking to solve problems, focusing on making conjectures and generalisations, and explaining and justifying conclusions using appropriate language.

Using and applying mathematics in real-world contexts

This involves pupils engaging in challenges that require them to use and apply their mathematical knowledge and skills in open-ended, real-world contexts.

This book is intended as a 'dip-in' resource, where teachers choose which of the three different types of challenge they wish pupils to undertake. A challenge may form the basis of part of or an entire mathematics lesson. The challenges can also be used in a similar way to the weekly bank of 'Learning activities' found in the *Busy Ant Maths* Teacher's Guide. It is recommended that pupils have equal experience of all three types of challenge during the course of a term.

The 'Solving mathematical problems' and 'Reasoning mathematically' challenges are organised under the different topics (domains) of the 2014 National Curriculum for Mathematics. This is to make it easier for teachers to choose a challenge that corresponds to the topic they are currently teaching, thereby providing an opportunity for pupils to practise their pure mathematical knowledge and skills in a problem solving context. These challenges are designed to be completed during the course of a lesson.

The 'Using and applying mathematics in real-world contexts' challenges have not been organised by topic. The very nature of this type of challenge means that pupils are drawing on their mathematical knowledge and skills from several topics in order to investigate challenges arising from the real world. In many cases these challenges will require pupils to work on them for an extended period, such as over the course of several lessons, a week or during a particular unit of work. An indication of which topics each of these challenges covers can be found on page 5.

Briefing

As with other similar teaching and learning resources, pupils will engage more fully with each challenge if the teacher introduces and discusses the challenge with the pupils. This includes reading through the challenge with the pupils, checking prerequisites for learning, ensuring understanding and clarifying any misconceptions.

Working collaboratively

The challenges can be undertaken by individuals, pairs or groups of pupils, however they will be enhanced greatly if pupils are able to work together in pairs or groups. By working collaboratively, pupils are more likely to develop their problem solving, communicating and reasoning skills.

You will need

All of the challenges require pupils to use pencil and paper. Giving pupils a large sheet of paper, such as A3 or A2, allows them to feel free to work out the results and record their thinking in ways that are appropriate to them. It also enables pupils to work together better in pairs or as a group, and provides them with an excellent prompt to use when sharing and discussing their work with others.

An important problem solving skill is to be able to identify not only the mathematics, but also what resources to use. For this reason, many of the challenges do not name the specific resources that are needed.

Characters

The characters on the right are the teacher and the four children who appear in some of the challenges in this book.

Mrs Patel

Sabah

Fabien

Joseph

Lucy

Think about ...

All challenges include prompting questions that provide both a springboard and a means of assisting pupils in accessing and working through the challenge.

What if?

The challenges also include an extension or variation that allows pupils to think more deeply about the challenge and to further develop their thinking skills.

When you've finished, ...

At the bottom of each challenge, pupils are instructed to turn to page 80 and to find a partner or another pair or group. This page offers a structure and set of questions intended to provide pupils with an opportunity to share their results and discuss their methods, strategies and mathematical reasoning.

When you've finished, turn to page 80.

Solutions

Where appropriate, the solutions to the challenges in this book can be found at *Busy Ant Maths* on Collins Connect and on our website: collins.co.uk/busyantmaths.

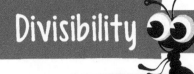
Challenge

If a number is divisible by 4, the last two digits of the number are divisible by 4.

If a number is divisible by 3, the sum of its digits is divisible by 3.

Are Fabien and Sabah right?

Why? Why not?

Think about ...

Check by using a selection of 2-digit and 3-digit numbers.

Make sure you provide examples to explain your decisions and your rules.

What if?

Think of a rule for numbers divisible by 6.

What about a rule for numbers divisible by 9?

When you've finished, turn to page 80.

Ordering numbers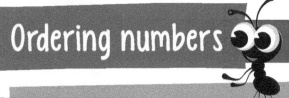

Challenge

| −6 | | −3 | | 0 | | 4 |

Look at the cards above.

Write numbers for each of the blank cards so that the seven numbers are in order.

How many different ways can you think of?

Think about ...

Work systematically to find all the different ways. Which numbers will stay the same and which numbers will change?

Recording your results carefully will help you to spot any patterns.

What if?

What if you use these numbers instead?

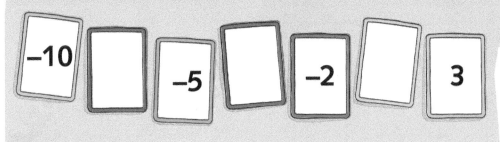

| −10 | | −5 | | −2 | | 3 |

When you've finished, turn to page 80.

Challenge

Investigate which numbers round to 350 when rounded to the nearest 10 **and** 400 when rounded to the nearest 100.

Which numbers round to 6250 when rounded to the nearest 10 **and** 6200 when rounded to the nearest 100?

Think about ...

Remember, your numbers must round to **both** the nearest 10 **and** 100, and in 'What if?' to **both** the nearest 100 **and** 1000.

Think carefully about the largest and smallest numbers you think will fit, and then try the numbers 1 more and 1 less to check.

What if?

What if the numbers round to 6600 when rounded to the nearest 100 **and** 7000 when rounded to the nearest 1000?

When you've finished, turn to page 80.

Making numbers

Challenge

Shuffle a set of 1–9 digit cards.

Deal the top four cards and place them face up on the table.

Use these digit cards to make 24 different 4-digit numbers.

Order your numbers from smallest to largest.

Then round each number to the nearest multiple of 10.

You will need:
- set of 1–9 digit cards

Think about ...

Record your work systematically to help you find the 24 different 4-digit numbers.

Think about representing your ordered and rounded numbers on number lines.

What if?

Round each of your 24 numbers to the nearest multiple of 100.

Round each number to the nearest multiple of 1000.

When you've finished, turn to page 80.

Dice calculations

Challenge

Roll a 1–6 dice four times.

After each roll of the dice, write down the number so you don't forget it.

Use the four numbers rolled to make 24 different 3-digit numbers.

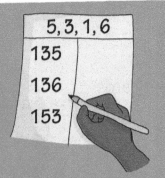

5, 3, 1, 6
135
136
153

You will need:
- 1–6 dice
- 0–9 dice or set of 0–9 digit cards

Which two of your 3-digit numbers have:

- the smallest total?
- the smallest difference?

- the greatest total?
- the greatest difference?

Think about ...

Working systematically will help you figure out the 24 different 3-digit numbers and 4-digit numbers.

Think carefully when identifying pairs of numbers with the smallest and greatest differences.

What if?

It is also possible to make 24 different 4-digit numbers using the four digits you rolled. Which pairs of 4-digit numbers have:

- the smallest total?
- the smallest difference?
- the greatest total?
- the greatest difference?

What if you use a 0–9 dice or a set of 0–9 digit cards instead of a 1–6 dice?

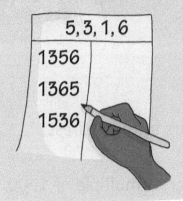

5, 3, 1, 6
1356
1365
1536

When you've finished, turn to page 80.

Challenge

Step 1
Write down a 3-digit number where each digit is different.

Step 2
Reverse the digits in the number.

Step 3
Subtract the smaller number from the larger number.

Step 5
Add that number to the answer of the subtraction.

Step 4
Reverse the digits in the answer.

Repeat the above steps, starting with different 3-digit numbers.

What do you notice?

Think about ...

Remember, each digit in your 3-digit number must be different, so you can't have, for example, the numbers 211 or 484.

At each step, you need to be working with 3-digit numbers.

What if?

What happens when you choose a 3-digit number where the ones and hundreds digits only have a difference of 1, such as **867**, **425**, **392** or **718**? What do you need to do to still end up with the same result?

When you've finished, turn to page 80.

Challenge

Write down your phone number.

Separate the digits and find the total.

Investigate separating the digits in other ways, finding the total each time.

How many different totals can you make?

What is the greatest total?

What is the smallest total?

My phone number is 01632 960289.

(01632) 960289

$1 + 6 + 3 + 2 + 9 + 6 + 0 + 2 + 8 + 9 = 46$

$16 + 32 + 96 + 02 + 89 = 235$

$163 + 2960 + 289 = 3412$

$1632 + 9602 + 89 = 11\,323$

Think about ...

Don't use the zero at the start of the area code.

Only use the digits to make 1-, 2-, 3- or 4-digit numbers – don't use the digits to make numbers greater than 9999.

What if?

What if you ignore the area code and separate the digits to make subtraction calculations? However, you must make sure that your answers are always positive numbers.

(01632) 960289

$960 - 289 = 671$

$9602 - 89 = 9513$

$96 - 02 - 89 = 5$

When you've finished, turn to page 80.

Challenge

Find a partner.

Using a set of dominoes:

- remove these two dominoes from the set as you won't be using them:

- take these five dominoes, lay them face down in a group and mix them up:

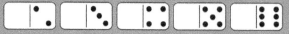

- take the remaining 21 dominoes, lay them face down in a separate group and mix them up.

You will need:
- set of 6 × 6 dominoes

Each person chooses one domino from each group.

Lay your two dominoes underneath each other similar to this:

'Read' your dominoes as a 2-digit number and a 1-digit number, and multiply the two numbers together.

The person with the greatest product wins that round.

Who wins more rounds after five turns each?

$$\begin{array}{r} 3\ 5 \\ \times\quad 4 \\ \hline 1\ 4\ 0 \\ \scriptstyle 2 \end{array}$$

Think about ...

After each go, place the domino that shows a blank and a number (for example) back in the group with the other similar dominoes and mix up the dominoes.

Think carefully about how you arrange your dominoes to help you make the greatest product possible.

What if?

Work with your partner to find pairs of dominoes that make a division calculation without any remainders. Here's one to get you started:

$21 \div 3 = 7$

When you've finished, turn to page 80.

Challenge

You will need:
- set of 0–9 digit cards

Choose five digit cards to complete the multiplication calculation below.

☐☐ × ☐ = ☐☐

In each calculation, you can only use each digit card once.

How many different multiplication calculations can you make?

Think about ...

When you've found one calculation, think about how you can use it to make other calculations.

How can you use your multiplication calculations to help you find division calculations?

What if?

What if you choose four cards to complete the division calculation below?

☐☐ ÷ ☐ = ☐

What if you choose five cards to complete this division calculation?

☐☐ ÷ ☐ = ☐☐

When you've finished, turn to page 80.

Challenge

32×5

Fabien worked out the answer to 32×5 like this:

$$
\begin{array}{r}
3\,2 \\
\times\quad 5 \\
\hline
1\,6\,0 \\
\hline
{\scriptstyle 1}
\end{array}
$$

Sabah worked out the answer using this method:

32×5

$4 \times 8 \times 5$

$4 \times 40 = 160$

Show how you would use Sabah's method to work out the answers to these calculations:

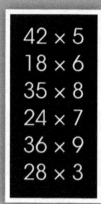

42×5
18×6
35×8
24×7
36×9
28×3

Think about ...

Use factors and your knowledge of the multiplication tables up to 12×12.

Use the commutative law to help you, for example: $6 \times 4 = 4 \times 6$ or $5 \times 9 \times 6 = 5 \times 6 \times 9$.

What if?

Check your answers, using either Fabien's method or a different method.

Write two multiplication calculations where Sabah's method is a good method to use.

Now write two other multiplication calculations where Sabah's method isn't the best method to use. Can you explain why?

When you've finished, turn to page 80.

Fractions of amounts

Challenge

0   2   3   4   5   6   7   8   9

You will need:
- set of 0–9 digit cards

Choose four of the remaining digit cards to complete the fraction statement.

$\frac{1}{\square} \times \square\square = \square$

In each statement, you can only use each digit card once.

What other different fraction statements can you make?

Think about ...

Use what you know about the multiplication and division facts for the multiplication tables up to 12×12 to help you.

When you've found one fraction statement, think about how you can use it to make other statements.

What if?

What if you choose five of the remaining digit cards to complete this fraction statement?

$\frac{1}{\square} \times \square\square = \square\square$

What if you choose five digit cards to complete this fraction statement?

$\frac{\square}{\square} \times \square\square = \square$

What if you choose six digit cards to complete this fraction statement?

$\frac{\square}{\square} \times \square\square = \square\square$

When you've finished, turn to page 80.

Making decimals

Challenge

Shuffle a set of 1–9 digit cards.

Deal the top three cards and place them face up on the table.

How many different decimals can you make using these digit cards?

Order your decimals from smallest to largest.

Then round each decimal with 1 decimal place to the nearest whole number.

You will need:
• set of 1–9 digit cards

Think about ...

Record your work systematically to help you find decimals with 1 and 2 decimal places.

Think about representing your ordered and rounded decimals on number lines.

What if?

Round each decimal with 2 decimal places to the nearest whole number.

What statements can you make using the < and > signs to compare pairs of decimals?

When you've finished, turn to page 80.

Challenge

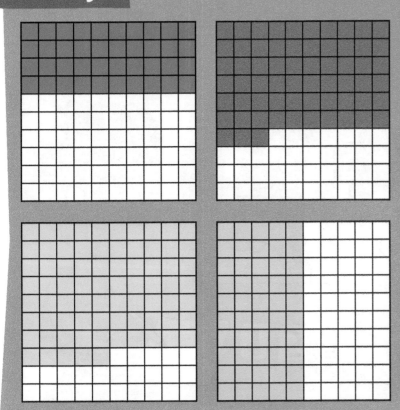

You will need:
- small blank 100 squares
- coloured pencils

How many different ways can you express the coloured squares in each of these 100 squares as a fraction and as a decimal?

Think about ...

Think about equivalent fractions.

Think about tenths and hundredths and their equivalents.

What if?

How many different ways can you express the white squares in each of the 100 squares as a fraction and a decimal?

Look at all the fractions and decimals for the orange square – those that describe the coloured squares and those that describe the white squares. What do you notice? Is this the same for the three other 100 squares?

When you've finished, turn to page 80.

Shopping basket

Challenge

Investigate different total masses you could have in your basket if you only bought three of the following items.

BUTTER $\frac{1}{4}$ kg

PASTA SAUCE 0·75 kg

COFFEE 200 g

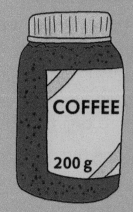

230 g GINGER BISCUITS

TUNA 160 g

JAM 370 g

Write each total in three ways:

- in kilograms and grams
- in grams only
- in kilograms only, using decimals.

YOGHURT $\frac{1}{2}$ kg

FLOUR 1·5 kg

Think about ...

$2\frac{1}{2}$ kg can also be written as 2 kg 500 g, 2500 g and 2·5 kg.

Think about when it is best to convert different units of measure into the same unit.

What if?

What if you bought four items?

What about five items?

When you've finished, turn to page 80.

Challenge

Make a drink using two different types of fruit juice. Choose the two fruits you think will taste good together.

Make the drink so that it is $\frac{7}{10}$ of one juice and $\frac{3}{10}$ of the other.

How many millilitres of each juice do you need to fill a small glass?

What about a medium glass?

What about a large glass?

200 ml

300 ml

500 ml

Think about ...

Think about equivalent fractions.

Remember that each small glass must contain 200 ml, each medium glass 300 ml, and each large glass 500 ml of fruit juice.

What if?

Now do the same thing, using three different types of juice.

Make the drink so that it is $\frac{5}{10}$ of the first juice, $\frac{3}{10}$ of the second juice and $\frac{2}{10}$ of the third juice.

What if you make a drink using all four different flavours of juice:

- $\frac{2}{5}$ of the third juice
- $\frac{1}{4}$ of the first juice
- $\frac{1}{4}$ of the second juice
- $\frac{1}{10}$ of the fourth juice?

When you've finished, turn to page 80.

Same and different

Challenge

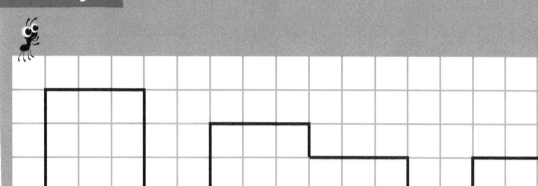

Roll a 1–6 dice twice.

Multiply together the two numbers rolled.

Draw a shape using the answer as the area.

Can you draw other shapes with the same area?

What is the perimeter of each of your shapes?

$4 \times 3 = 12\,cm^2$

Think about ...

If you roll a 1, roll the dice again.

Draw your shapes as accurately as possible.

What if?

Investigate drawing shapes that have areas of 24 cm² but different perimeters.

Investigate drawing shapes that have perimeters of 24 cm but different areas.

When you've finished, turn to page 80.

Challenge

The display on a digital clock uses light bars to create digits.

All the digits 0 to 9 can be made using up to seven light bars.

Make all the digits 0 to 9 using light bars.

Write the following times as they would appear on a 12-hour digital clock:

- when you woke up this morning
- the time school starts
- the time it is now
- when you have lunch
- the time school finishes
- when you go to bed.

Now write the times as they would appear on a 24-hour digital clock.

Think about ...

For the 12-hour digital clock, you will need to indicate what time of the day your clock is showing.

You don't need to write your times using light bars!

What if?

Lucy says: I go to bed at 8:30 p.m. most nights. If I had a 12-hour Roman numeral digital clock, this is what the time would look like.

VIII : XXX PM

Lucy also says: If I had a 24-hour Roman numeral digital clock, 8:30 p.m. would look like this.

XX : XXX

What if you write each of the times above as they might appear on a 12-hour and 24-hour Roman numeral digital clock? Remember, zero does not have its own Roman numeral, so you'll need to use the Hindu-Arabic numeral: 0.

When you've finished, turn to page 80.

Finding totals

Challenge

Choose different pairs of items and work out the total.

How many different combinations can you make?

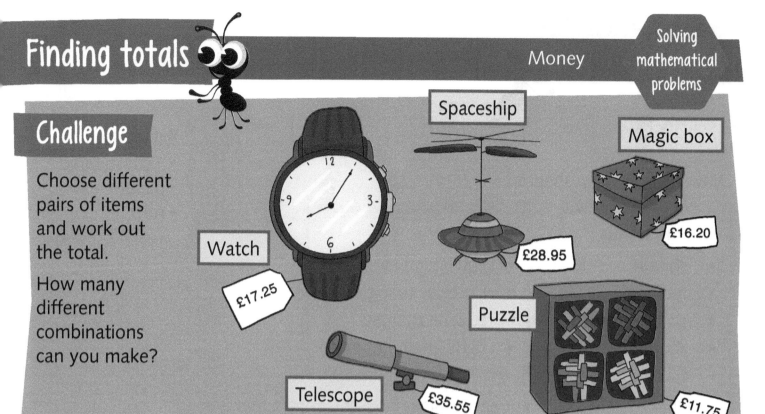

Spaceship

Magic box

£16.20

Watch

£17.25

£28.95

Puzzle

Telescope

£35.55

£11.75

Think about ...

Look carefully at each pair of items before you work out the total. Which totals can you work out mentally? Which totals do you need to use a written method for?

For the second 'What if?' question below, when working out what notes and coins the cashier might give you in change, use the smallest number of coins possible.

What if?

Look at your different totals. Using the chart below, work out the change you would receive for buying each pair of items.

Total of pair of items	£30 or less	£30.01 to £40	£40.01 to £50	£50.01 to £60	£60.01 to £70
Notes given to the cashier to pay for the items	£20 note and £10 note	two £20 notes	£50 note	£50 note and £10 note	£50 note and £20 note

When you've finished, turn to page 80.

What notes and coins might the cashier give you in change?

25

Challenge

Draw the following shapes.

You will need:
- 1 cm squared paper
- ruler

Quadrilaterals
- square
- rectangle
- parallelogram
- rhombus
- trapezium
- kite

Triangles
- equilateral triangle
- isosceles triangle
- scalene triangle
- right-angled triangle

For each shape, label the angles that are:
- acute angles
- right angles
- obtuse angles.

Think about ...

Draw your shapes as accurately as possible.

Think about how you're going to label the angles in each shape.

What if?

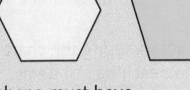

What do you notice about the types of angle in these regular shapes?

Draw each of the shapes below. Each shape must have at least one right angle, one acute angle and one obtuse angle. Label all the angles in each shape.

- irregular pentagon
- irregular hexagon
- irregular octagon

When you've finished, turn to page 80.

Symmetrical shapes

Challenge

Draw a 6 by 6 square on squared paper.

Draw the horizontal line of symmetry as shown.

Place 6 cubes above the horizontal line of symmetry to make a shape.

Use another 6 cubes to show the reflection of the shape in the mirror line.

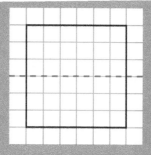

Record your completed shapes.

Make different shapes using between 5 and 10 cubes. Ask a friend to show the reflection of the shape in the horizontal line of symmetry.

Record the completed shapes.

You will need:
- about 30 centicubes and 1 cm squared paper or 30 interlocking cubes and 2 cm squared paper
- coloured pencils
- a friend

Think about ...

Make sure that your shapes also have colour symmetry.

Make shapes that touch the line of symmetry and shapes that do not.

What if?

What if you draw a vertical line of symmetry?

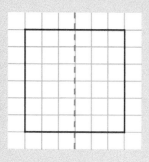

What about a diagonal line of symmetry?

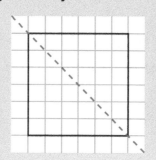

When you've finished, turn to page 80.

Grid route

Challenge

Use either a first quadrant coordinates grid with vertical and horizontal axes marked to 10, or draw and label a grid as shown here.

Mark the following points on the grid:

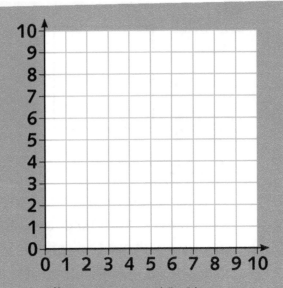

✗ red cross at (9, 5)   ✗ yellow cross at (6, 2)

✗ blue cross at (1, 7)   ✗ orange cross at (5, 9)

✗ green cross at (2, 3)

You will need:
- first quadrant coordinates grid (vertical and horizontal axes marked to 10) or squared paper
- ruler
- coloured pencils

Write a route to describe how to move from the green cross to the blue cross, passing through the orange cross.

Write a different route describing how to move from the green cross to the blue cross, passing through the orange cross.

Think about ...

Make sure that you have plotted the points correctly.

Make sure that your routes are the easiest routes possible, making the fewest turns.

What if?

Write a route describing how to move from the red cross to the yellow cross, passing though all the other coloured crosses.

Can you write a different route?

When you've finished, turn to page 80.

Challenge

The dots on the grid are the vertices of six different quadrilaterals.

The vertices of each quadrilateral are shown by a different coloured dot. The black dots are vertices of more than one quadrilateral.

Copy the grid and join up the dots to draw the six different quadrilaterals.

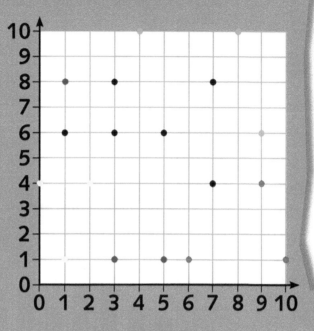

You will need:
- several copies of first quadrant coordinates grids (vertical and horizontal axes marked to 10) or squared paper
- ruler
- coloured pencils

Think about ...

A quadrilateral is a polygon with 4 straight sides, 4 vertices and 4 angles.

Can you name the six most common quadrilaterals?

What if?

Draw the six quadrilaterals on a different coordinates grid with some of the shapes sharing a vertex.

Now redraw the grid. This time, draw one similar to the grid above, but just show the vertices of the six quadrilaterals – each quadrilateral with different coloured dots to represent the vertices, and shared vertices shown by black dots.

When you've finished, turn to page 80.

Give this grid to a friend. Can they join up the dots to draw the six different quadrilaterals?

Interpreting the weather

Statistics

Solving mathematical problems

Challenge

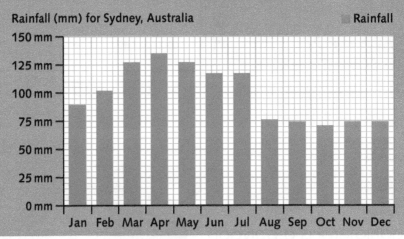

Average minimum and maximum temperatures (°C) for Sydney, Australia

— Max temp
— Min temp

Look at these two graphs. The line graph shows the average minimum and maximum temperatures for Sydney, Australia. The bar chart shows Sydney's average rainfall.

What statements can you make about Sydney's average temperatures and rainfall?

Rainfall (mm) for Sydney, Australia — Rainfall

Think about ...

Make statements describing and comparing month-by-month temperatures and rainfall, and also between the different seasons.

Can you describe what the weather might be like on a typical day in a particular month?

What if?

Look at the table on page 31 showing the average minimum and maximum temperatures for London, England, and London's average rainfall. What statements can you make comparing temperatures and rainfall in Sydney and London?

When you've finished, turn to page 80.

30

Challenge

Using the data in the table below, draw a line graph showing the average minimum and maximum temperatures for London, England, and a bar chart showing London's average rainfall.

You will need:
- 1 cm squared paper or graph paper
- ruler
- coloured pencils

Look at the two graphs on page 30 to help you.

London, England

Jan	Feb	Mar	Apr	May	Jun	Jul	Aug	Sep	Oct	Nov	Dec
Average minimum temperature (°C)											
2	2	3	6	8	12	14	13	11	8	5	4
Average maximum temperature (°C)											
6	7	10	13	17	20	22	21	19	14	10	7
Average rainfall (mm)											
54	40	37	37	46	45	57	59	49	57	64	48

What statements can you make about London's average temperatures and rainfall?

Think about ...

Think carefully about how you're going to label the vertical axes on your graphs. What labelled and unlabelled intervals are you going to use?

For the 'What if?', think about how you can show both the temperature and rainfall scales on one graph. What labelled and unlabelled intervals are you going to use?

What if?

Look at your two graphs. Draw **one** graph that shows London's average minimum and maximum temperatures **and** average rainfall.

When you've finished, turn to page 80.

Challenge

The numbers 15, 30, 45 and 60 are all multiples of 3 and 5.

Multiples of 3	15, 30, 45, 60	Multiples of 5
3, 6, 9, 12, 18, 21, 24, 27, 33, 36, 39, 42, 48, 51, 54, 57		5, 10, 20, 25, 35, 40, 50, 55

Think about multiples of 2, 3, 4, 5, 6, 7, 8, 9 and 10.

Draw Venn diagrams to show which numbers are common multiples of two different numbers.

Write about what you notice about the common multiples.

Think about ...

Before you choose which two (and, for the 'What if?', three) numbers have multiples in common, think carefully about which numbers are likely to have lots of common multiples.

Make sure you include up to the 12th multiple for each number. If you can, write more than 12 multiples.

What if?

Joseph also says:

60 is a multiple of 3, 4 and 5.

Multiples of 3: 3, 6, 9, 18, 21, 27, 33, 39, 42, 51, 54, 57

15, 30, 45

Multiples of 5: 5, 10, 25, 35, 50, 55

60

12, 24, 36, 48

20, 40

Multiples of 4: 4, 8, 16, 28, 32, 44, 52, 56

Which numbers are common multiples of three different numbers?

Draw Venn diagrams to justify your answers.

Write about what you notice about the common multiples.

When you've finished, turn to page 80.

Challenge

"I have two more place value counters in my hand."

"The only 4-digit number you can make using all the counters on the table and the two counters in your hand is 3213."

Is Joseph right?

Explain why.

Think about ...

Think about the different combinations of two counters Lucy may be holding.

Working systematically will help you identify all the numbers possible.

What if?

Joseph also says:

Is Joseph right?

Explain why.

Using just some of the seven place value counters on the table, you can make 18 numbers less than 1000.

When you've finished, turn to page 80.

Challenge

How many different 4-digit numbers can you write where the tens digit is 6 and all four digits add up to 9?

Write your numbers in order, smallest to largest.

$$\boxed{} + \boxed{} + \boxed{6} + \boxed{} = 9$$

Think about ...

Work systematically to help you identify any patterns and help you find all of the numbers.

What do the thousands, hundreds and ones digits need to add up to?

What if?

You can make 15 different 4-digit numbers where the tens digit is 4 and the four digits add up to 9.

$$\boxed{} + \boxed{} + \boxed{4} + \boxed{} = 9$$

Write the 15 numbers in order, starting with the smallest.

When you've finished, turn to page 80.

Challenge

The largest whole number that rounds to 600 is 599.

The smallest whole number that rounds to 3000 is 2050.

Are Sabah and Fabien correct?

Explain why.

Think about ...

Think about whether you are rounding to the nearest multiple of 10, 100 or 1000. Does this affect which is the smallest or largest whole number?

Think carefully about which is the most important digit when rounding to the nearest 10, 100 or 1000 and whether you need to round up or down.

What if?

Round 597 to the nearest 10.

Round 597 to the nearest 100.

What do you notice?

What other numbers round in this way?

What pattern do you notice?

When you've finished, turn to page 80.

Challenge

Copy and complete the calculation.

How did you work out what the missing digits were?

There is more than one solution. How many different solutions can you find?

Explain how you worked out the different solutions.

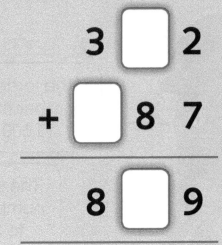

Think about ...

Once you've found two solutions, compare the calculations. Look for any patterns to help you find other solutions.

Make sure you check your answers.

What if?

How many different solutions can you find for this calculation?

Explain how you worked out the missing digits and the different solutions.

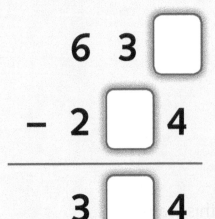

When you've finished, turn to page 80.

Challenge

Estimate which of these calculations have an answer between 500 and 600.

374 + 285	209 + 278	758 – 286
863 – 329	261 + 311	924 – 306

Explain your decisions.

Estimate which of these calculations have an answer between 6000 and 6500.

4630 + 1605	3500 + 2920	7200 – 900
8400 – 2600	1630 + 4320	9350 – 2565

Explain your decisions.

For which of the 12 calculations did you find it easy to make an estimate of the answer?

Which calculations did you find more difficult?

Can you explain why?

Think about ...

Don't work out the answers to begin with – just make an estimate.

Think about how rounding numbers will help you with your estimations.

What if?

Look at the calculations where you have estimated the answers are between 500 and 600, and between 6000 and 6500. Work out the answers.

Now use a different calculation to check your answers.

When you've finished, turn to page 80.

Challenge

Write down two 3-digit even numbers.

If you were to add the two numbers together, would your answer be an odd or an even number?

What if you were to find the difference between the two numbers? Would the answer be an odd or an even number?

Estimate the answers to the addition and subtraction calculations. Calculate the answers. Were your predictions correct? Why? Why not?

What happens when you write down two 3-digit odd numbers? **683 167**

What about one odd and one even 3-digit number? **549 238**

Think about ...

Use your mental strategies for adding and subtracting pairs of 2-digit numbers to help you make your predictions for 3-digit numbers.

After you have estimated and calculated your answers, check them. How are you going to do this?

What if?

Sabah says:

Are these good predictions? Explain why.

If I add 378 and 454, I know that the answer will be 700 and something because 300 add 400 is 700.

I also know that if I subtract 378 from 454, the answer will be 100 and something because 400 subtract 300 is 100.

When you've finished, turn to page 80.

Challenge

Find the matching pairs of calculations.

Explain why each pair of calculations is a match.

5 × 5 × 7

4 × 8 × 9

18 × 7

25 × 7

3 × 9 × 8

16 × 8

4 × 4 × 8

32 × 9

9 × 2 × 7

27 × 8

Think about ...

Remember the **commutative law** for multiplication: that the order in which the numbers are multiplied doesn't matter.

Remember, any two numbers that make a larger number when multiplied together are called **factor pairs**. So 4 and 6 are a factor pair of 24, and so too are 2 and 12, and 3 and 8.

What if?

For each of these calculations, write a matching calculation to create pairs of calculations similar to those above.

36 × 7

6 × 7 × 8

3 × 8 × 6

18 × 9

Compare your pairs of calculations with a friend.

Are your pairs of calculations the same?

If they're different, who's right?

Could you both be right?

Explain why.

When you've finished, turn to page 80.

Challenge

Fabien, Lucy, Sabah and Joseph were all given the same calculation to answer. Here is how they each worked out the correct answer.

Fabien

					2	6	4
				×			8
8	×	4	=			3	2
8	×	6	0	=		4	8	0
8	×	2	0	0	=	1	6	0	0	+
					2	1	1	2
					1	1	

$8 \times 4 = 32$
$8 \times 60 = 480$
$8 \times 200 = 1600$
$+$
2112

Joseph

	2	6	4
×			8
		3	2
	4	8	0
1	6	0	0	+
2	1	1	2
1	1	

Lucy

×	200	60	4
8	1600	480	32	= 2112

Sabah

		2	6	4
×	5	3	8
	2	1	1	2

Which of these methods would you choose to solve a similar calculation? Why?

Which of the methods above would you never use? Why?

Would you use a different method? If so, what is it?

Think about ...

Which method gives you the answer the quickest? Which method is the most consistent for getting the correct answer?

Your preferred method might not be exactly the same as one of the methods above. How is it different?

What if?

Work out the answer to 743 × 6 using your preferred method.

Then check your answer using a different method.

When you've finished, turn to page 80.

Challenge

Replace each letter with a digit from 1 to 9.

Identical letters must be replaced by the same digit.

The same digit cannot be used for more than one letter.

$$
\begin{array}{r}
A\ B \\
\times\qquad C \\
\hline
B\ C\ D \\
\hline
\end{array}
\qquad
\begin{array}{r}
E\ F\ B \\
\times\qquad G \\
\hline
H\ F\ I\ F \\
\hline
\end{array}
$$

If B = 3 and F = 8, what is the value of each of the other letters?

Explain how you worked out the value of the remaining seven letters.

Think about ...

Is it possible to have more than one solution to any of the calculations?

Once you've worked out the value of one of the letters, use trial and improvement to help you work out the value of the remaining letters.

What if?

$$Q \times R \times S = T\ T\ Q$$

If R = 7, what are the values of Q, S and T?

When you've finished, turn to page 80.

Challenge

Sort these fraction calculations into two groups: those with the right answer and those with the wrong answer.

Explain your reasoning as to why each calculation is right or wrong.

$$\frac{4}{9} + \frac{3}{9} = \frac{7}{9}$$

$$\frac{3}{10} + \frac{5}{10} = \frac{8}{20}$$

$$\frac{3}{5} \times 20 = 12$$

$$\frac{11}{12} - \frac{3}{12} = \frac{7}{12}$$

$$\frac{2}{3} \times 18 = 9$$

$$\frac{5}{8} - \frac{3}{8} = \frac{1}{4}$$

$$\frac{5}{8} + \frac{1}{8} = \frac{3}{4}$$

Think about ...

Remember the rules for adding, subtracting and multiplying with fractions.

Think about equivalent fractions.

What if?

For each wrong calculation, work out the right answer.

Show all your working.

Choose any two calculations. Draw a diagram to represent each calculation.

When you've finished, turn to page 80.

Challenge

Dividing a 1-digit number by 100 gives an answer with 2 decimal places.

Dividing a 2-digit number by 100 gives an answer with 2 decimal places.

Dividing a 1-digit number by 10 gives an answer with 1 decimal place.

Dividing a 2-digit number by 10 gives an answer with 1 decimal place.

Are Sabah's, Lucy's, Fabien's and Joseph's statements always true, sometimes true or never true?

Provide examples to explain why.

Think about ...

Make sure you include a range of different numbers when justifying your decisions.

What happens to the value of the digits when you divide a number by 10 or 100?

What if?

Fabien also says:

Is Fabien correct?

Explain your reasoning.

When you order a set of numbers with 1 decimal place, the number with the largest number of tenths is always the largest number.

When you've finished, turn to page 80.

Challenge

Write a decimal number that lies between each pair of labelled numbers and fractions on these number lines.

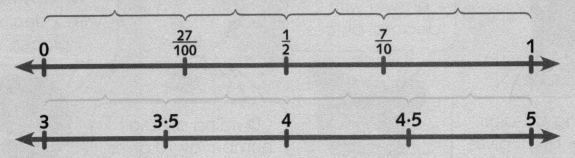

Can you think of a second decimal number that lies between each pair of labelled numbers and fractions?

Think about ...

Are you going to write a decimal with 1 or 2 decimal places?

Think about equivalent fractions and decimals, and equivalent tenths and hundredths.

What if?

Write a fraction that lies between each pair of labelled numbers on these number lines.

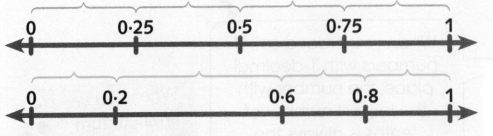

Can you write a second fraction that lies between each pair of labelled numbers?

When you've finished, turn to page 80.

Challenge

Is each statement true or false?

1000 ml = 1 l

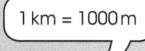
1 km = 1000 m

250 ml = $\frac{1}{2}$ l

2·25 l = 2250 ml

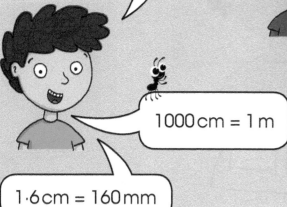

1000 cm = 1 m

1·6 cm = 160 mm

1 kg = 100 g

1·3 kg = 130 g

750 g = $\frac{3}{4}$ kg

Explain your reasoning.

Think about ...

What does the prefix 'kilo' refer to? What about 'centi' and 'milli'?

Think about the relationship between kilo…, centi… and milli… .

What if?

Order these lengths, starting with the shortest.

$\frac{1}{4}$ of 800 cm $\frac{1}{2}$ of 3 m $\frac{3}{4}$ m   120 cm

Explain your reasoning.

When you've finished, turn to page 80.

Challenge

The perimeter of a rectangle is 18 cm.

Draw the rectangle.

Can you draw a different rectangle with a perimeter of 18 cm?

The perimeter of a square is 20 cm.

Draw the square.

Can you draw a different square with a perimeter of 20 cm?

You will need:
- 1 cm squared paper
- ruler

Think about ...

What is the same about a square and a rectangle? What is different about a square and a rectangle?

Draw as many different rectangles and squares as you can.

What if?

The area of a rectangle is 24 cm².

Draw the rectangle.

Can you draw a different rectangle with an area of 24 cm².

The area of a square is 25 cm².

Draw the square.

Can you draw a different square with an area of 25 cm²?

When you've finished, turn to page 80.

Challenge

On a 24-hour digital clock, there are times when the minutes digits add up to the hours digits.

2 + 2 = 4

4 + 9 = 13

Are there more of these times between midnight (00:00) and noon (12:00) or between noon (12:00) and midnight (00:00)?

Make a prediction, and then find out.

Think about ...

Be systematic when writing the times – this will help you spot any patterns.

Make sure you write about how you arrived at your predictions and what you did to find out the answers.

What if?

How many times on a 12-hour digital clock does the sum of the hours and minutes digits total 12?

Predict, and then find out.

1 + 1 + 4 + 6 = 12

How many times on a 24-hour digital clock does the sum of the hours and minutes digits total 24?

Predict, and then find out.

When you've finished, turn to page 80.

Money | Reasoning mathematically

Challenge

Would you rather have:
- 16 × 20p coins
- 7 × 50p coins
- 3 × £1 coins?

Would you rather have:
- $\frac{1}{2}$ of £13
- $\frac{1}{4}$ of £22
- $\frac{3}{4}$ of £8?

Would you rather have:
- 525p
- £5.60
- 620p?

Would you rather have:
- £1, £2 and 2 × 20p
- £5 and 7 × 10p
- $\frac{1}{2}$ of £15?

For each child, which amount would you rather have?

Explain your thinking.

Think about ...

Think carefully about which operation to use to work out the total of each amount.

What if?

Write each of the 12 amounts above using pounds and pence, then order the amounts, starting with the least.

When you've finished, turn to page 80.

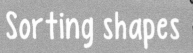

Sorting shapes

Properties of shapes

Reasoning mathematically

Challenge

Sort and label these shapes into different groups.

How many different ways can you sort the shapes?

Think about ...

Think carefully about the different properties of 2-D shapes and the criteria you're going to use to sort the shapes. What is the same and what is different about the shapes?

It is probably best to show how you have sorted the shapes by writing the names of the shapes rather than drawing them.

What if?

Sort, but do not label, the shapes using a different criterion.

Show your sorting diagram to a friend. Can they work out the criterion you used to sort the shapes and label your groups?

When you've finished, turn to page 80.

49

Challenge

Who drew which angle?

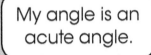

a   b   c   d

You will need:
- 1 cm squared paper
- ruler

I drew a right angle.

My angle is an acute angle.

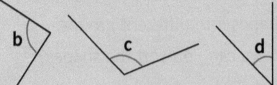

I drew the smallest angle.

My angle is an obtuse angle.

Explain your thinking.

Think about ...

What do you need to consider when identifying the name of an angle?

For the 'What if?' think carefully about the properties of the different quadrilaterals.

What if?

The diagonals of which of these quadrilaterals meet at right angles?

- square
- rectangle

- parallelogram
- rhombus

Prove it.

When you've finished, turn to page 80.

Challenge

Two of these shapes have been reflected correctly and two have not.

Identify the two shapes that have **not** been reflected correctly.

Explain why they are incorrect. Then draw the correct reflections.

You will need:
- 1 cm squared paper
- ruler
- coloured pencils

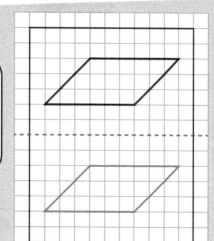

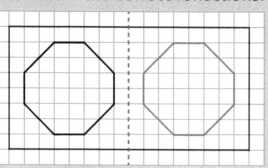

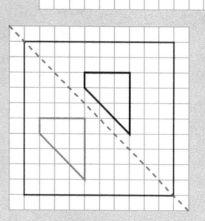

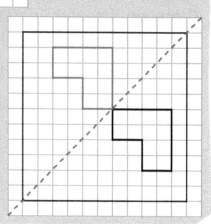

Think about ...

Think carefully about how you're going to explain why a shape has not been reflected correctly.

In the four diagrams above, the original shape is drawn in black and the reflection is drawn in grey.

What if?

Copy these patterns and colour extra squares to make them symmetrical.

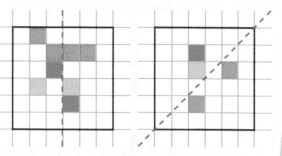

When you've finished, turn to page 80.

Which pattern did you find harder to complete? Explain why?

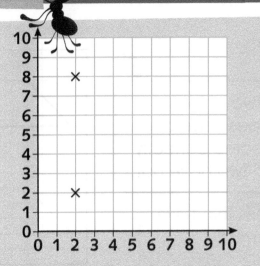

Challenge

Use either a first quadrant coordinates grid with vertical and horizontal axes marked to 10, or draw and label a grid as shown.

Mark the points (2, 2) and (2, 8) on the grid.

These are the coordinates of two vertices of a quadrilateral.

What might the quadrilateral be?

What could be the coordinates of the other two vertices of the quadrilateral?

You will need:
• first quadrant coordinates grid (vertical and horizontal axes marked to 10) or squared paper
• ruler

How many different quadrilaterals can you identify with the coordinates of two vertices of each quadrilateral (2, 2) and (2, 8)?

Write the set of coordinates for the four vertices of each of your quadrilaterals and name the quadrilateral.

Think about ...

Remember that a quadrilateral is a polygon with 4 straight sides, 4 vertices and 4 angles.

You don't have to draw the quadrilaterals, but you can if it helps you to visualise what the quadrilaterals might be and what the coordinates of their vertices are.

What if?

A square and a rectangle are two quadrilaterals. There are at least four other types of quadrilateral you should be able to identify.

Draw and name at least one of these quadrilaterals. Remember, two of the vertices of the quadrilateral must be (2, 2) and (2, 8).

When you've finished, turn to page 80.

Challenge

No. There are two different ways.

No! You're both wrong – there are more than two different ways.

There's only one way to walk from the school 🏫 to the shop 🏪.

Who's right?

Prove it.

Use words such as: **left**, **right**, **up** and **down** to describe routes between various points on the island.

Also, describe your routes using coordinates.

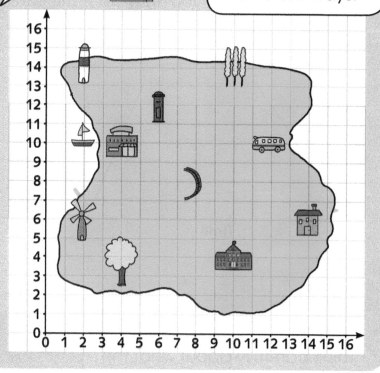

Think about ...

Try to describe the shortest routes possible.

If you need to cross the river you need to use the bridge ⌒.

What if?

Use coordinates to describe how you would walk from:

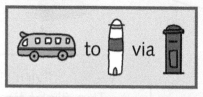 to via

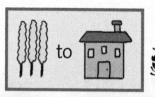 to

 to via

 to

Write a set of coordinates for a route between two or three different places on the island. Give your route to a friend. Can they describe where you're going from and to?

When you've finished, turn to page 80.

53

Challenge

What could each of these graphs be showing?

What conclusions can you draw from the data in the graphs?

You will need:
- 1 cm squared paper or graph paper
- ruler

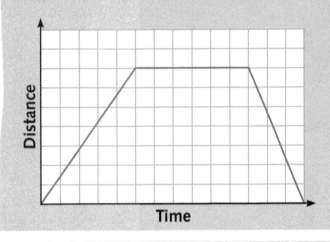

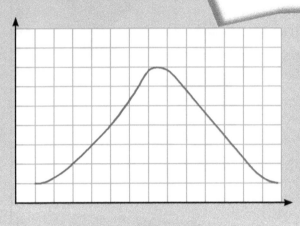

Think about ...

Think carefully about what the horizontal and, particularly, the vertical axes in the second graph might be.

Can you think of two slightly different stories for each graph?

What if?

Draw a time graph to show the following:

We are flying from London to Nice. The flight time is 2 hours. We will remain in Nice for just $\frac{1}{2}$ an hour while passengers disembark the plane and new passengers board the plane for the return flight to London.

When you've finished, turn to page 80.

Challenge

Class 4 collected data about the birds they spotted in the nature area of their school.

They presented their results in three different ways.

What's the same about these presentations of the data? What's different?

Which do you find is the easiest to read and interpret? Why is this?

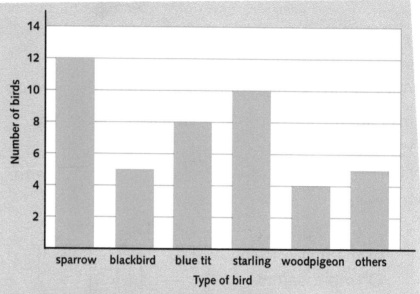

Type of bird	Number of birds
sparrow	12
blackbird	5
blue tit	8
starling	10
woodpigeon	4
others	5

$= 2$

Think about ...

What are the strengths of each of the three different types of data presentation?

What are the drawbacks?

What if?

What statements can you make comparing the different types of bird spotted in the nature area?

When you've finished, turn to page 80.

Ancient Greek numbers

Challenge

Ancient Greeks did not use the digits 1–9 as we do today.

Instead, the letters of their alphabet were used with ' or , to represent numbers.

Use the Ancient Greek alphabet below to write the following dates:

- 776 BC (the year of the first Ancient Olympic Games)
- 395 AD (the year of the last Ancient Olympic Games)
- 1896 AD (the year of the first Modern Olympic Games).

Greek alphabet

α'	β'	γ'	δ'	ε'	ϛ'	ζ'	η'	θ'
1	2	3	4	5	6	7	8	9
ι'	κ'	λ'	μ'	ν'	ξ'	ο'	π'	φ'
10	20	30	40	50	60	70	80	90
ρ'	σ'	τ'	υ'	φ'	χ'	ψ'	ω'	⅄'
100	200	300	400	500	600	700	800	900
,α	,β	,γ	,δ	,ε	,ϛ	,ζ	,η	,θ
1000	2000	3000	4000	5000	6000	7000	8000	9000

Think about ...

Use these three examples to help you:
31 = λα'
752 = ψνβ'
8193 = ,ηρφγ

What similarities do you notice between the Ancient Greek way of writing numbers and the Hindu-Arabic numeral system we use today? What differences are there between the two systems?

What if?

Choose ten different 2-digit numbers and write them using Ancient Greek numbers.

Now choose ten different 3-digit and 4-digit numbers.

When you've finished, turn to page 80.

Ancient Egyptian number symbols

Challenge

The Ancient Egyptians had a number system using seven different symbols.

Use the Ancient Egyptian number symbols to write:

- 2560 BC (the year the Great Pyramid at Giza was built)
- 1343 BC (the year of Tutankhamun's death)
- 69 BC (the year of Cleopatra's birth).

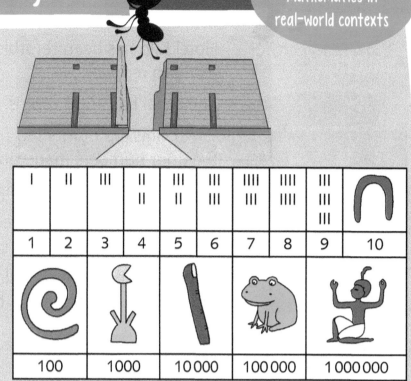

I	II	III	II II	III II	III III	IIII III	IIII IIII	III III III	∩
1	2	3	4	5	6	7	8	9	10
100		1000		10 000		100 000		1 000 000	

Think about ...

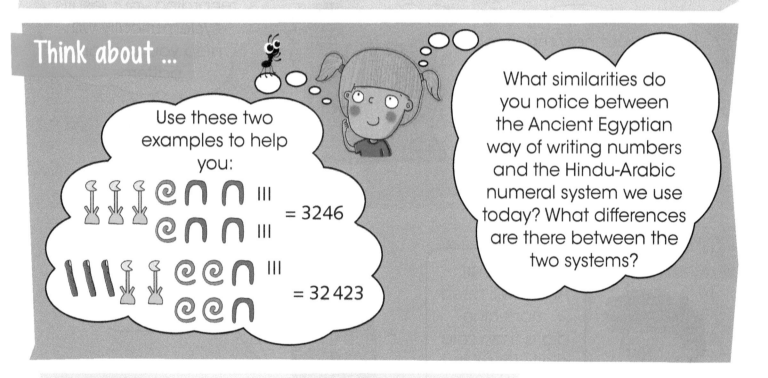

Use these two examples to help you:

= 3246

= 32 423

What similarities do you notice between the Ancient Egyptian way of writing numbers and the Hindu-Arabic numeral system we use today? What differences are there between the two systems?

What if?

Choose ten different 2-digit numbers and write them using Ancient Egyptian number symbols.

Now choose ten different 3-digit and 4-digit numbers.

When you've finished, turn to page 80.

Hotel Kilo

Challenge

Hotel Kilo has been refurbished and is almost ready to reopen.

The hotel has 1000 rooms.

The rooms are numbered from 1 to 1000, with the door numbers made out of brass digits.

So Room 476 needs the digits: 4 7 6

Albert, the handyman, needs to screw all the numbers onto all the doors.

How many brass zeros will Albert need for all the room numbers from 1 to 1000?

Think about ...

Think about how many zeros (and nines) there are from 1 to 100. How can this help you to work out how many zeros (and nines) there are for the next 100 numbers?

Working and recording your results systematically will help you spot any patterns.

What if?

I think that Albert will need more nine digits than zero digits for all the room numbers from 1 to 1000.

I think Albert will need fewer nines than zeros.

Who's right?

How many more or fewer nines than zeros will Albert need?

When you've finished, turn to page 80.

Dice imaginings

Challenge

Place four 1–6 dice side by side to make a wall.

Without looking, work out how many spots there are on the bottom of the wall.

Without looking, work out how many spots there are altogether where the faces of the four dice touch.

Can you arrange the dice so that there is the same number of spots on the top of the wall as on the bottom of the wall? Can you explain why?

You will need:
• four 1–6 dice

Think about ...

Quickly make each wall with the dice and don't move the dice. Work out the number of spots you can't see.

Think carefully about how many numbers you need to add or multiply together where the faces of the dice touch.

What if?

What if you use three dice?

Without looking, what is the product of the sets of spots on the bottom of the wall?

Without looking, what is the product of the sets of spots where the faces of the three dice touch?

When you've finished, turn to page 80.

Streets

Challenge

Estimate how many people live in your street.

Think about the number of houses and blocks of flats in your street.

Think about how many people live in each house and flat.

What fraction do you think are adults? What fraction do you think are children?

Write about how you arrived at your estimates.

Think about ...

Remember your aim is to make 'accurate' estimates.

For the 'What if?', think carefully about the different categories to use when classifying the street.

What if?

Choose a busy street near your school.

How is the space on either side of the street used?

Think about the different categories you can classify the street into.

What fraction of the street is used for each of your categories?

When you've finished, turn to page 80.

Number plate calculations

Challenge

K211 PKY

GK37 AXJ

BH84 HWU

KT03 ERS

211
37
84
3

Write down the number on each of the number plates in your school car park.

Choosing any three of these numbers, how close can you get to 100 using one or more of the four operations: + − × ÷?

Think about ...

You may decide to just use addition, or addition and subtraction. For example:
84 + 3 + 37
or
84 + 37 − 3

If you decide to use multiplication or division as well, make sure that you multiply or divide before you add or subtract. For example:
37 + 84 ÷ 3 ✗
84 ÷ 3 + 37 ✓

What if?

What if you use any four of the numbers?

When you've finished, turn to page 80.

Carleton Harvest Festival

Using and applying mathematics in real-world contexts

Challenge

Every October, Carleton has a Harvest Festival.

This year, 140 people have already bought tickets to the festival.

The organisers of the festival have two kinds of table. One kind of table seats 6 people, the other kind seats 8 people.

How many tables of each kind will the organisers need to use to seat all 140 people?

Show your working.

Think about ...

The organisers have at least 30 of each size of table to use if they need to. The organisers don't want to have any spare seats at a table – they want full tables of 6 or 8 people.

There is more than one possible combination, using 6-seater and 8-seater tables. How many different combinations can you find?

What if?

Three days before the festival, an additional 30 people buy tickets to the festival.

How many tables of each kind will the organisers need to use to seat all 170 people?

When you've finished, turn to page 80.

Intercity travelling

Challenge

Kilometre chart	Birmingham	Cardiff	Liverpool	London
Birmingham		185	156	193
Cardiff	185		333	247
Liverpool	156	333		357
London	193	247	357	

If you wanted to visit each of the four cities above by car, what is the shortest route you could take?

How long is this route?

The distance between London and Liverpool is 357 km.

Think about ...

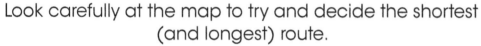

Look carefully at the map to try and decide the shortest (and longest) route.
You can start your journey from any of the four cities.
You're not making a 'round trip' back to where you started, you're just travelling to the four cities, for example:
London → Birmingham → Cardiff → Liverpool ✓
London → Birmingham → Cardiff → Liverpool → London ✗

What if?

What is the longest route?

When you've finished, turn to page 80.

Tides

Challenge

Choose a coastal location.

Look up information about tides in the local newspaper or online.

What patterns can you spot between the different tides?

Write statements comparing the times and heights of tides over the course of a week.

You will need:
- weather section from a newspaper showing tide tables

	21.07.2017		22.07.2017		23.07.2017	
	Time	Height (m)	Time	Height (m)	Time	Height (m)
Low	02:42	0·9	03:30	0·8	04:12	0·8
High	08:39	2·6	09:25	2·7	10:09	2·8
Low	15:14	0·5	16:02	0·3	16:45	0·3
High	21:23	2·6	22:07	2·6	22:49	2·5

Think about ...

Make comparisons between tides during a 12-hour period.

Make comparisons between tides during a 24-hour period.

Make comparisons between tides from one day to the next.

What if?

Choose a different coastal location.

What are the similarities and differences between tides in the two locations?

When you've finished, turn to page 80.

Money trail

Challenge

Approximately how many 10p coins would you need to make a trail the length of your classroom?

How much money would your trail be worth?

Write about how you made your estimate.

You will need:
• measuring equipment
• one coin of each denomination

Think about ...

Remember:
• You are only expected to make an 'accurate' estimate.
• For 'What if?', you can only use one coin of each denomination to find out how many coins you would need and how much the trail would be worth.

You will need to do some measuring and also some rounding.

What if?

What if you use £1, £2, 50p, 20p or 5p coins instead?

How many coins would you need?

How much would each trail be worth?

When you've finished, turn to page 80.

Visiting your local area

Challenge

Imagine you work for the local tourist board.

Plan an itinerary for a day's visit to your local area.

What places of interest should you include?

Approximately how long will it take people to get from one place to another?

How long would you expect someone to spend at each place?

Don't forget to allow time for them to eat and rest!

You will need:
- brochures about places of interest in the local area

Think about ...

Think about how you're going to present your itinerary. You might want to include a map.

Your timings need only be approximations.

What if?

Some of the places you include in your itinerary may charge an entry fee.

What is the expected cost of your day's visit to your local area?

You should also think about the cost of providing lunch and maybe even a morning and/or afternoon snack!

What other expenses are there likely to be during the day?

When you've finished, turn to page 80.

Going on holiday

Challenge

Choose a place for you and your family to go on holiday.
It might be a holiday at home or abroad.

How are you going to get there?

How far away is it?

How long will it take you to travel there?

Investigate how much it will cost for your family to go on holiday for a week.

You will need:
- holiday brochures

Think about ...

Think about:
- accommodation
- travel costs
- food
- other expenses.

What if?

You can make all the arrangements for a holiday yourself, or you can go on a package holiday where everything is organised for you by a travel agent or company, with arrangements for transport, accommodation, and sometimes food, made at an inclusive price.

Can you find a package holiday similar to the holiday you have arranged for your family?

Which holiday gives better value for money? Why do you think this is?

When you've finished, turn to page 80.

School recycling and rubbish

Using and applying mathematics in real-world contexts

Challenge

What things can you recycle?

What things does your school recycle?

Approximately how much of each of these things does your school recycle each week?

You will need:
- measuring equipment

Think about ...

How are you going to measure and record each amount?

Remember, you are only expected to make an 'accurate' estimate.

Think about how many recycling and rubbish bins there are in your school, how big they are, and how often they're emptied.

What if?

Approximately how much rubbish does your school produce, which is not recycled, each week?

When you've finished, turn to page 80.

Challenge

Work out how much of each ingredient you would need to make a Spanish omelette for:

- 2 people
- 6 people
- 8 people
- 12 people.

Show all your working.

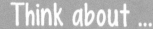

Spanish omelette
Serves 4

800 g sliced potatoes
60 ml olive oil
6 eggs
150 g chopped onions
2 garlic cloves
400 g tinned tomatoes
1 teaspoon chilli powder

Think about ...

Think about the best way to present your results so that you don't have to write out the list of ingredients lots of times.

For some of the ingredients, you will need to round up or round down to a suitable amount.

Think about the relationship between the different number of people you are cooking for and the number of servings each recipe is for. How can you use this relationship to work out the different quantities needed of each ingredient?

What if?

What if you made chocolate mousse for dessert?

How much of each ingredient would you need for 2, 4, 6 or 12 people?

Show your working.

Chocolate mousse
Serves 8

180 g dark chocolate
120 g butter
4 eggs
2 tablespoons of coffee

Work out approximately how much of each ingredient you would need to give everyone in your class a slice of Spanish omelette and a portion of chocolate mousse.

It wouldn't matter if there was some of each food left over – but not too much!

Make sure you show all your working.

When you've finished, turn to page 80.

Balanced menu

Challenge

Design a nutritious and balanced menu for a main meal.

Use the Food Pyramid to help you design your menu.

How much of each food group will you need to feed a family of four?

Work out the approximate cost of your menu.

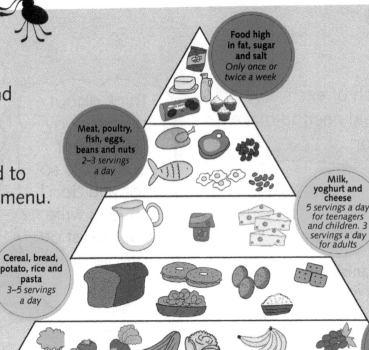

Food high in fat, sugar and salt
Only once or twice a week

Meat, poultry, fish, eggs, beans and nuts
2–3 servings a day

Milk, yoghurt and cheese
5 servings a day for teenagers and children. 3 servings a day for adults

Cereal, bread, potato, rice and pasta
3–5 servings a day

Vegetables, salad and fruit
5–7 servings a day

Drink at least 8 glasses of liquid a day – ideally water

You will need:
- different types of food packaging
- access to online supermarkets

Think about ...

Think about the portion sizes you are going to give everyone. How are the portions of each food group going to vary between children and adults?

Think about the recommended portion sizes on the Food Pyramid. Can you also include the approximate weight or volume of each food group?

What if?

Use the Food Pyramid to design an unhealthy, unbalanced menu for a main meal.

How much of each food group will you need to feed a family of four?

What is the approximate cost of your menu?

Which menu is more expensive? Why do you think this is?

When you've finished, turn to page 80.

Healthy and unhealthy food

Challenge

Investigate the nutrition information on the side or back of different types of food packaging.

Classify each food item into one of two groups: Healthy food or Unhealthy food.

Explain why you classified each type of food the way you did.

You will need:
• different types of food packaging

Think about ...

Think about how much fat, the amount of sugar and how many calories each food contains.

Refer to the Food Pyramid on page 70 to help you classify each food item.

What if?

Nutritional information is also displayed on the front of pre-packed food. These labels provide information on the number of grams of fat, sugars and salt, and the amount of energy (in kJ and kcal) in a serving of the food.

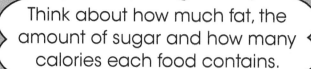

Energy	Fat	Saturated	Sugars	Salt
1047 kJ 251 kcal	4.0g	1.5g	35g	1g
	LOW	LOW	HIGH	MED
14%	5%	8%	39%	16%

Some foods use 'traffic light labels' that tell you if the food has low, medium or high amounts of fat, sugars and salt.

Green means low. The more green on the label, the healthier the food.

Amber means medium. You can eat foods with all or mostly amber on the label most of the time.

Red means high. These are the foods you should eat less often and in small amounts.

Use the information on the front of food packaging to classify each food item as **Healthy food** or **Unhealthy food**. Explain why you classified each type of food the way you did.

Compare your classifying of the side or back of different food packaging with the information on the front of pre-packed food. Is there any difference? If so, why? If not, why not?

When you've finished, turn to page 80.

71

Timetables

Challenge

Draw a daily timetable showing the starting and finishing times for each of the films at a local cinema.

Write the starting and finishing times using 12-hour notation and a.m. and p.m.

You will need:
- access to online cinema times

Then draw a weekly timetable. Write the starting and finishing times using 24-hour notation.

Think about ...

Think about which day the cinema changes its programme and which day of the week you are going to start your timetable.

Think about how your daily and weekly timetables will differ.

Make sure that your timetables are as easy to read as possible.

What if?

Design a timetable of your daily life.

On it you should show how you spend a week during term time.

Don't forget to include the time you spend eating, sleeping, watching TV, playing, ...

When you've finished, turn to page 80.

School holidays

Challenge

Draw a calendar for the next academic year.

Mark which days are school days and which days are holidays.

You will need:
- list of next year's school term dates
- ruler

Think about ...

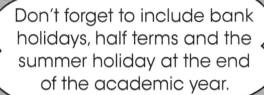

Don't forget to include bank holidays, half terms and the summer holiday at the end of the academic year.

Think carefully about what you need to do and the best order in which you should do these things.

What if?

We come to school for more than half the number of days in a year.

No we don't! We come to school for less than half the year.

Given that there are normally 365 days in a year, do you come to school for more or less than half the year?

When you've finished, turn to page 80.

Who's right? Explain why.

73

How fit?

Challenge

Work with a friend.

One of you skips for 2 minutes.

Once the person skipping has finished:

- work out how quickly they're breathing
- work out how fast their heart is beating.

Pause and take a rest!

Now the same person runs on the spot for 2 minutes.

- work out how quickly they're breathing
- work out how fast their heart is beating.

Swap roles and repeat the activities.

Each of you then compares your results for skipping and running.

Then compare your results with each other.

Write about your results.

You will need:
- clock or watch with a seconds hand or a stopwatch
- skipping rope

Think about ...

Think about how you're going to measure how quickly you're breathing and how fast your heart is beating.

In your results, write about how fit you are.

What if?

When you're both finished, find another pair and compare all your results.

Write about your results.

When you've finished, turn to page 80.

Challenge

Work with a partner or in a small group.

Plan a meal out with some friends.

At the end of the meal, share the bill evenly between you.

Mario's Pizza

PIZZA	Small (one person)	Large (two people)
Hawaiian	£6.50	£10.80
Margherita	£4.20	£8.50
Vegetarian	£5.30	£9.80
Meat feast	£7.80	£12.00
Four seasons	£5.40	£9.90

CREATE YOUR OWN
Each extra topping: small: 80p; large: £1.20

cheese, tomato, olives, salami, chicken, mushrooms, pineapple, peppers, onion, chilli

EXTRAS
Garlic bread
£1.80 per person

Salad
small: £3.20; large: £5.80

DRINKS
Water: £1
Fruit juice: £3
Can of soft drink: 80p
1·5 litre bottle of soft drink: £2

DESSERTS
Chocolate cake: £2.70
Apple pie: £2
Toffee delight: £3.10
Ice cream: £2.60
(Chocolate, Cookie, Vanilla)

Think about ...

Is it cheaper to buy a large pizza and share the cost or buy two small pizzas?

How are you going to work out what each person needs to pay? Make sure that it's fair for everyone.

What if?

What if each person could spend no more than £10 on a meal?

Your meal must include at least a pizza, drink and dessert.

When you've finished, turn to page 80.

Angle time

Challenge

Excluding 12 o'clock and 6 o'clock, investigate for which 'o'clock' times the angle between the minute and hour hands is:

- an acute angle
- a right angle
- an obtuse angle.

Then order the ten 'o'clock' times according to the size of the angles, starting with the smallest angle.

Think about ...

You are thinking about the smaller angle for each time, for example:

this angle

not this angle

Think carefully about ordering the times according to the size of the angles. Are there any angles that are the same size? If so, how are you going to show the order of these times?

What if?

Write a half past time where the angle between the minute and hour hands is:

- an acute angle
- an obtuse angle.

What about a quarter past time where the angle between the minute and hour hands is:

- an acute angle?
- an obtuse angle?

What about a quarter to time where the angle between the minute and hour hands is:

- an acute angle?
- an obtuse angle?

When you've finished, turn to page 80.

Street map

Challenge

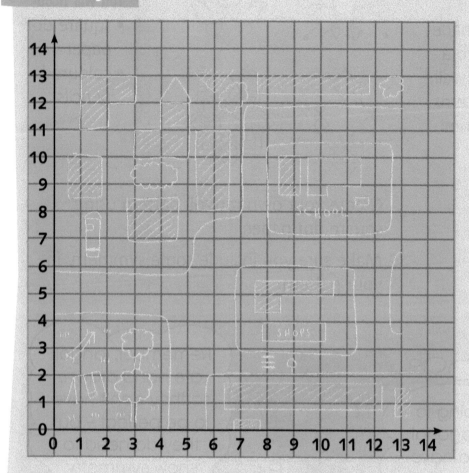

You will need:
- squared paper
- ruler
- coloured pencils

Using squared paper, draw a map of the area around your school.

Mark on your map at least ten different things in the area, for example: the school, a shop, a post box, your house, a large tree ...

Label the axes: 0, 1, 2, 3, 4, ...

Now make a list of the different things on your map, with their grid reference.

Think about ...

Start by marking in the roads and then adding the different features.

Think about the scale of your map. Make sure that the different features on your map are in the correct position and that their sizes are in proportion to each other.

What if?

Write directions for someone to find their way from one place on your map to another.

When you've finished, turn to page 80.

Rangoli

Challenge

A rangoli is a colourful design made on the floor near the entrance to a home to welcome guests. At Diwali, Hindus draw bright rangoli patterns to encourage the goddess Lakshmi to enter their homes.

Look at these four rangoli patterns. They all show examples of symmetry.

They have each been drawn using square dot paper.

You will need:
- square dot paper
- coloured pencils

Create your own rangoli pattern using square dot paper.

Make sure you include some symmetry in your pattern.

Think about ...

Is your line of symmetry going to be vertical, horizontal or diagonal, and is your rangoli going to have one or two lines of symmetry?

Remember, rangoli are hand drawn, so they don't have to be perfect – but they do need to show symmetry.

What if?

Look at these rangoli patterns. They all use colour but still have symmetry.

Create your own rangoli pattern using colour.

When you've finished, turn to page 80.

Festivals

Challenge

Work with a partner or in a small group.

Make a list of all the different festivals that you know about.

Try to find out when each festival occurs. Is it at the same time each year or does it vary from year to year?

As a group, decide which you think are the six most popular festivals.

Then ask at least 30 people which of these six festivals is their favourite.

Organise and present your findings as a bar chart, pictogram or other graph.

You will need:
- 1 cm squared paper or graph paper
- ruler
- coloured pencils

Think about ...

Think about local and national festivals, as well as religious and non-religious festivals.

Decide which is the best way to present your findings.

What if?

Different festivals celebrate in a similar way, for example using lights or fireworks, fasting, or exchanging gifts.

Look back at your list of different festivals. Investigate which festivals have celebrations in common.

Draw a chart or diagram showing the similarities between festival celebrations.

When you've finished, turn to page 80.

Share Share your results.

Discuss Discuss any results that are different.

Which result is correct?

Might there be more than one solution?

Share Share the methods used.

Discuss Discuss the similarities and differences in the methods used.

Which method worked best?

Are there any other ways to go about solving the problem?

Share Share what you have learned.

Discuss Discuss what you would do the same, and what you would do differently next time.

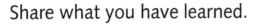

Is there anything you would do differently?

What have you learned for next time?

What would you do the same?